Everyday Science Experiments in the Playground

John Daniel Hartzog

The Rosen Publishing Group's
PowerKids Press™
New York

Some of the experiments in this book were designed for a child to do together with an adult.

Published in 2000 by The Rosen Publishing Group, Inc.
29 East 21st Street, New York, NY 10010

Photo Credits and Photo Illustrations: p. 4 © Skjold Photography; pp. 6, 7, 8, 12, 14, 15, 16, 17, 19, 20 by Shalhevet Moshe; p. 11 CORBIS/Dean Conger.

First Edition

Book Design: Michael de Guzman

Hartzog, John Daniel
     Everyday science experiments in the playground / by John Daniel Hartzog.
        p.   cm. — (Science surprises)
     Includes index.
     Summary: Suggests activities and experiments that demonstrate the principles of science at work in a playground, exploring such topics as gravity, centrifugal force, and shadows.
     ISBN 0-8239-5457-9 (lib. bdg.)
     1. Science—Experiments Juvenile literature. [1. Science—Experiments. 2. Experiments.] I. Title. II. Series: Hartzog, John Daniel. Science surprises.
Q164.H272   1999
507'.8—dc21   99-18291
                                                                            CIP

Manufactured in the United States of America

Contents

Discovery at the Playground

The playground is one of everybody's favorite places to play. Did you know that it is also a great place to learn about science? Almost all of the fun things at the playground rely on the laws of science. Going down the slide would not be as much fun without **friction** or **acceleration**. The merry-go-round would be boring without **centrifugal force**. Understanding how things work can make them more fun.

◀ *The playground is a great place to learn about science.*

Why Things Fall

When you start to climb a ladder or cross the monkey bars do you know what would happen if you let go? You would fall down, of course. That's because of a **force** called **gravity**. Let's do an **experiment** to learn more about gravity.

Climb to the highest safe place in your playground. From there, drop two balls that are the same size at the same time. They will both hit the ground at the same time. Now drop one large ball and one small ball. The big ball is heavier

Materials Needed:
- 2 small balls (or rocks)
- 1 large ball (or rock)

than the small ball. Which one do you think will hit the ground first? They hit the ground at the same time because gravity pulls on all things with the same strength.

Even though the basketball is bigger than the tennis ball, they both hit the ground at the same time.

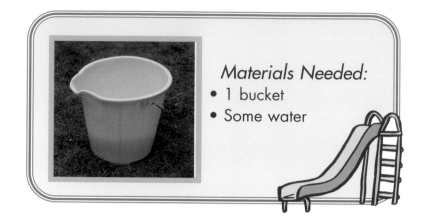

Materials Needed:
- 1 bucket
- Some water

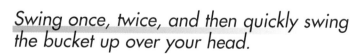

Swing once, twice, and then quickly swing the bucket up over your head.

If you are swinging fast enough, centrifugal force will hold the water in the bucket. ▶

8

The Merry-Go-Round

Does your playground have a merry-go-round? If so, you know that when you ride it you have to hold on at all times or you will get thrown off. The force that would make you fall is called centrifugal force. Centrifugal force pulls spinning things towards the outside of the circle they're spinning in. Try this experiment to help you understand how it works.

Hold a half-full bucket of water by your side in one hand. Quickly swing the bucket up and over your head in a circle. If you swing fast enough, you can swing the bucket around as many times as you want and the water won't come out. Were you surprised that you did not get wet? The water was held in the bucket by centrifugal force.

The Seesaw

Seesaws are fun because you can go up high, come back down, and go up again without having to work very hard. A seesaw is easy to ride because it is a good **model** of a machine called a **lever**. Levers help us to lift things. Let's see how much we can lift on the seesaw.

A seesaw usually only works if the two people riding it are the same weight. Otherwise the heavier person would stay down and the lighter person would stay up. Instead, there needs to be a **balance**. One way to balance two people who are not the same weight is to have the heavier person move toward the middle of the seesaw. The closer to the middle that person is, the less they pull their side down. Then riding on the seesaw can be easy again!

You don't have to work very hard to go up and down on a seesaw.

Spinning Without Getting Dizzy

When you spin around do you get dizzy? Wouldn't it be fun to be able to spin without getting dizzy? Try this experiment and you will learn how.

Standing straight, pick a spot straight ahead of you to fix your eyes on. Now start spinning, but keep your eyes on that spot. To do that, you'll have to keep your head still more of the time. Just turn it at the very last moment, when you start to feel a pull in your neck. Then quickly find your spot again. You may be spinning, but your head won't be.

◀ *Pick a spot to look at while you spin and you won't get dizzy.*

Telling Time with the Shadows

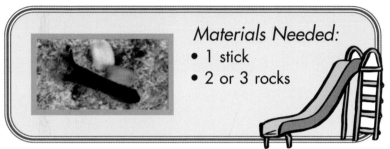

Materials Needed:
- 1 stick
- 2 or 3 rocks

Have you ever gotten in trouble for staying at the playground longer than you were supposed to? It is hard to remember to check the time when you are having fun. What if telling time was more fun? Here's a way to make telling the time a playground game.

Take a straight stick and poke one end in the ground. Do you see a shadow? The shadow will be on the side of the stick that is farther away from the sun. Put a rock at the end of the shadow. Come back when it's time for you to leave. Where is the shadow now? It has moved away from the rock. Put a new rock at the end of the new shadow. If you set one rock for when you get to the playground and one for when you leave, you will always know when it's time to go home.

▲
As time passes, the shadow will move. Mark the times when you come and go with rocks, and then you can use the shadows to tell the time!

Grass Whistles

Playgrounds are fun because we get to holler, yell, and be loud. Your voice works because of a small strip of **cartilage** in your throat called a **larynx**. As air passes over this strip it **vibrates** and makes noise. Would you like to make a noisemaker that works like your larynx? Find a **blade** of grass that is flat, long, and wide. Place the bottom of the grass between the soft parts of your hand below your thumbs. Push these parts of your hands together so they hold on to the bottom of the grass. Pull the grass straight

Hold the grass between your hands.

Pull it tight with your index fingers.

up with your index fingers. Then, push the sides of the tips of your thumbs together to hold the grass tightly in place. There should be a small space between your thumbs where you can see the grass. Pucker your lips and blow through this hole. Blow hard, soft, and everything in between until you get the grass to make a good sound. Experiment with the looseness or tightness of the grass to get different sounds. Your larynx can tighten and loosen to create different sounds just like this blade of grass does.

Close your thumbs to hold the grass in place.

Blow through the hole between your thumbs.

Swing Magic

For some people, swings are the best part of the playground. If you watch carefully, you will see that swings follow a very unusual pattern. A swing takes the same amount of time to get from one end of its motion to the other, no matter how high or low it is swinging. That's because a swing is a good example of a **pendulum**. Pendulums can be found in many clocks because they measure the passage of time.

Find a friend and mark a line in the dirt under the swing. Have your friend get into the swing then step aside to watch. When she starts to swing, count how long it takes for her to come back over the line. No matter how high or low or fast or slow she is swinging, it will always take the same time to get back to that line. That is the magic of a pendulum.

Mark a line in the dirt under where the swing naturally hangs.

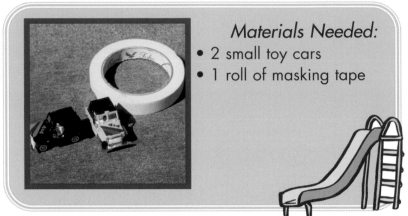

Materials Needed:
- 2 small toy cars
- 1 roll of masking tape

Wrap one of the cars in masking tape.

Send both cars down the slide at once.

The car without the tape creates less friction.

Ride the Slide

The slide is one of the fastest rides on the playground because of acceleration. Acceleration is the way objects speed up once they're in motion. The slide will get faster as you grow because acceleration increases with weight. Acceleration also depends on friction. Friction slows things down. Let's try an experiment to see how friction affects speed on a slide. Find two toy cars and wrap one in masking tape, covering the wheels. Hold the two cars at the top of the slide and let them go at the same time. Though the car with the tape will be a little heavier, the one with wheels creates less friction. Because the wheels move as the car goes down the slide, they don't rub against the slide like the masking tape does. The friction slowed the acceleration of the heavier car.

Science Works Everywhere

After trying all of these experiments, the playground will never seem the same again. A seesaw is now like a machine that helps to lift things, and a swing is like part of a clock. The science of the playground looks at how things move. This kind of science is called **physics**.

With science your playground has become more fun and interesting. If you look carefully enough you will find the ideas you have learned at the playground all around you. The magic of all science is that it can help you to see the whole world in a new and different way.

Glossary

acceleration (ek-SEL-er-ay-shun) Increasing in speed.

balance (BAL-uhns) To make equal in weight, amount, or force.

blade (BLAYD) A single leaf of grass.

cartilage (KAR-tuh-lij) A strong, flexible substance in people and animals that forms part of the body.

centrifugal force (sen-TRIF-ih-gul FORS) The force that pushes outward on a turning or spinning object.

experiment (ek-SPER-ih-ment) To test the effect of something.

force (FORS) Something that moves or pushes on something else. Force can change the direction or speed of a moving body.

friction (FRIK-shun) The rubbing of one thing against another. Friction makes it harder for things to move.

gravity (GRAV-ih-tee) The force that pulls things downward towards the center of the earth.

larynx (LAIR-inks) The top part of the windpipe where the vocal cords can be found.

lever (LEH-vuhr) A tool used to lift heavy things.

model (MOD-uhl) A copy of something that is smaller and simpler than the real thing.

pendulum (PEN-juh-luhm) A weight hung from a fixed point in such a way that it can swing back and forth in a very regular motion.

physics (FI-ziks) The scientific study of matter and energy and the laws which govern them. Physics is the study of motion, force, light, heat, sound, and electricity.

vibrates (VY-brayts) When an object moves back and forth very quickly.

Index

Web Sites:

You can find out more about playground science on this Web site:

http://www.ccslabs.com/backyard/